James P. Walker

The book of Raphael's Madonnas

James P. Walker

The book of Raphael's Madonnas

ISBN/EAN: 9783742860552

Manufactured in Europe, USA, Canada, Australia, Japa

Cover: Foto ©berggeist007 / pixelio.de

Manufactured and distributed by brebook publishing software
(www.brebook.com)

James P. Walker

The book of Raphael's Madonnas

BOOK

OF

Raphael's Madonnas.

BY JAMES P. WALKER.

NEW YORK:

LEAVITT AND ALLEN.

1860.

Contents.

Illustrations.

PHOTOGRAPHED BY E. HUFNAGEL.

PREFACE.

When the idea of preparing a photographically illustrated book of Raphael's Madonnas first suggested itself, it was accompanied by the natural desire to make the collection complete; i. e., to include photographs of all the "Holy Families," "Virgins," and "Madonnas," of this great master; that his matchless performances in this department might be made as familiar and easily comprehendible, as they have been made in others, through the publication of the "Book of Raphael's Cartoons," etc. But this end, desirable as it is, was manifestly impossible, from the circumstance that but a limited number of the original pictures have ever been reproduced by the engraver; and of those at any time engraved,—amounting in all to about thirty,—several are so rare as not to be obtainable in this country, or only very inferior copies of them. In view of these circumstances, it was deemed wisest to make a selection of the choicest and most universally-esteemed of these productions; with the purpose of issuing, at some future period, a Second Series, if the first volume meet the approbation of the public.

In selecting the engravings for photographing, care has been exercised to secure as true copies of the original pictures as could be found; a consideration which will be appreciated by those who are familiar with the liberty which competent engravers are accustomed to exercise, and the carelessness of the incompetent, in reproducing the work of any artist, especially one of the early masters.

It would be easy to point to engravings of Raphael's Madonnas, well executed mechanically, but in which the design of the Painter has been so altered by the engraver, as to raise a question in the mind of the beholder, which of the originals had been followed; the result being a sort of fancy sketch "founded on fact," and occupying, in Art, the anomalous position of "Historical Novels" in literature.

The illustrative sketches which accompany the photographs have been compiled from a great variety of fources, and it is hoped will be found to enhance the intereft of the collection. The authorities principally depended upon, are :—

QUATREMERE DE QUINCEY'S LIFE OF RAPHAEL,
VASARI'S LIVES OF THE PAINTERS, ETC.,
KUGLER'S HAND-BOOK OF PAINTING,
MRS. JAMESON'S LEGENDS OF THE MADONNA,
MRS. JAMESON'S SKETCHES OF ART,

though a great number of other works have been incidentally confulted; while the poetical literature of England and America has been gleaned to furnifh appropriate and agreeable accompaniments to the defcriptive fketches.

That the volume, notwithftanding its faults of execution, will not prove wholly unacceptable, we feel affured, from the fubject which it ftrives to illuftrate; "a fubject," in the words of a modern authorefs, beft qualified to difcourfe thereon, " fo confecrated by its antiquity, fo hallowed by its profound fignificance, fo endeared by its affociations with the fofteft and deepeft of our human fympathies, that the mind has never wearied of its repetition, nor the eye become fatiated with its beauty. Thofe who refufe to give it the honor due to a religious reprefentation, yet regard it with a tender, half-willing homage ; and when the glorified type of what is pureft, loftieft, holieft in womanhood, ftands before us, arrayed in all the majefty and beauty that accomplifhed Art, infpired by faith and love, could lend her, and bearing her divine Son, rather enthroned than fuftained on her maternal bofom, 'we look, and the heart is in heaven!'—and it is difficult, very difficult, to refrain from an ORA PRO NOBIS."

With thefe words of explanation, we commend the volume, which has afforded us many hours of delightful, if laborious occupation, in the preparation, to the cultivated and the tafteful.

J. P. W.

Boston, July, 1859.

Raphael.

WASHINGTON ALLSTON.

BY Heaven imprelfed with genius' feal,
  An eye to fee, and heart to feel,
  His foul through boundlefs nature roved,
  And feeing felt, and feeling loved.
But weak the power of mind at will
To give the hand the painter's fkill;
For mortal works, maturing flow,
From patient care and labor flow:
And, hence reftrained, his youthful hand
Obeyed a mafter's dull command;
But foon with health his fickly ftyle
From Leonardo learned to fmile;
And now from Buonarotti caught
A nobler form; and now it fought

2

Of color fair the magic spell,
And traced her to the Friar's* cell.
No foolish pride, no narrow rule,
Enslaved his soul; from every school,
Whatever fair, whatever grand,
His pencil like a potent wand,
Transfusing, bade his canvas grace.
Progressive thus, with giant pace,
And energy no toil could tame,
He climbed the rugged mount of Fame:
And soon had reached the summit bold,
When Death, who there delights to hold
His fatal watch, with envious blow
Quick hurled him to the shades below.

---

* Fra Bartolomeo.

Outline of Raphael's Life and Genius.

" In Raphael's hands, art performs its higheſt, and, indeed, its only legitimate function, it makes us better men."—HILLARD.

ITH no intention of preparing an elaborate Bio-graphical Sketch of the diſtinguiſhed maſter, whoſe wonderful compoſitions form the ſubject of this volume, it is fitting that the following ſketches and ſelections ſhould be preceded by a brief narrative of the principal events of that ſhort but brilliant career; and ſuch tributes to his ſurpaſſing genius from thoſe qualified to pronounce them, as may ſerve to illuſtrate and enforce his claim to precedence among the throng, whoſe productions crowd the galleries of the paſt, conſtituting at once their patent of nobility, and their crown of immortality.

RAPHAEL DE SANZIO (or RAFFAELLO, as Vaſari and the

modern Italians write it) was born in the small town of Urbino, in the Papal States, on Good Friday, March 24th, 1483. He received his first instruction in art from his father, Giovanni Santi, a painter of little reputation; and, in 1494, was placed under the tuition of Pietro Perugino, a master not unworthy his illustrious pupil. Here he remained for three years, when Perugino, being summoned by business to Florence, Raphael essayed trials of his powers, and made several excursions in the environs of Perugia.

In 1504, he removed to Florence, where he remained, with the exception of occasional visits to Perugia and Bologna, till 1508. In that year he was called to Rome by Pope Julius II., to assist in the adornment of the Vatican, a labor which occupied him, with numerous intermissions, several years. His house, built by himself, near the Piazza Vaticano, is still pointed out to visitors in the "Eternal City." Between 1512 and 1520, the majority of his matchless Madonnas, Holy Families, Portraits, etc., were executed: the Cartoons at Hampton Court were executed 1515-16; the frescoes of the Farnesina, 1518. Besides his labors in this department, he was employed, from 1515, in building the new Basilica of St. Peter, having that year been appointed by the Pope architect of that structure.

His death, which was fudden as it was untimely, is faid to have been caufed by a fever, induced by a fevere cold, contracted during a converfation with the Pope about the progrefs of St. Peter's; which took place in one of the vaft halls of the palace, whither Raphael, on receiving a fummons, had proceeded in fuch hafte as to arrive in a profufe perfpiration.

He expired on Good Friday, April 6th, 1520, at the age of 37.

After laying in ftate, at his own houfe, in the apartment where hung his laft work, the Transfiguration, his remains were conveyed, amidft the lamentations of the whole city, to the ancient Pantheon—the Church of Santa Maria de la Rotunda, and depofited, in accordance with his laft requeft, at the foot of the chapel he had endowed, where his fepulchre now is.

For more than a century the Academy of St. Luke, at Rome, exhibited, in a glafs cafe, a fkull, which it was pretended was that of Raphael; and the author of " Rome in the Nineteenth Century," alludes, in terms of becoming difguft, to the exhibition.   In 1833, to filence the queries which had arifen upon the fubject, the tomb of Raphael was opened with great care, and in the prefence of many of the higheft dignitaries of the Church and State; and, after its repofe of more than three centuries, the fkeleton of the great mafter was found entire.   A mould was taken of the fkull; and the fecond inhumation took place on the evening of the 18th of October, with great pomp, the interior of the Rotunda being funereally illuminated on the occafion.

All his biographers unite in afcribing to Raphael great beauty of perfon, and yet greater beauty of character.   Of agreeable manners, modeft, thoughtful of others, obliging, he difarmed the jealoufy which his extraordinary and verfatile genius and rapid advancement were calculated to infpire.   Indeed, Vafari affures us, "that he was never feen to go to Court but furrounded and accompanied, as he left his houfe, by fome fifty painters, all men of ability and diftinction, who attended him thus to give evidence of the honor in which they held him."

His mental acquirements were confiderable and refpectable.   That he did not lack for timely and fatirical wit, and boldnefs withal, the following anecdote will indicate :—

It is faid that while engaged in painting his celebrated frefcoes, he was vifited by two cardinals, who began to criticife his work, and found fault without underftanding it.

" The apoftle Paul has too red a face," faid one.

"He blufhes even in heaven to fee what hands the Church has fallen into," replied the indignant painter.

It is little to fay, coldly, that for invention, compofition, expreffion, and grace, Raphael far excelled all his predeceffors and contemporaries; while the univerfal teftimony of thofe familiar with his paintings, is, that they are pervaded by a namelefs charm, perceptible by all perfons of tafte, and diftinguifhing them from all other works of art, but rather to be felt, than analyzed and defcribed in fet terms.

Richardfon, in his Effays, as quoted by Hazlitt, after a rapid furvey of the peculiar excellences of the moft celebrated artifts, concludes thus:—"But ah ! the pleafure, when a connoiffeur and lover of art has before him a picture or drawing, of which he can fay, this is the hand, thefe are the thoughts of him (Raphael) who was one of the politeft, beft-natured gentlemen that ever was; beloved and affifted by the greateft wits and the greateft men then in Rome : of him who lived in great fame, honor, and magnificence, and died extremely lamented; miffed a cardinal's hat only by dying a few months too foon; but was particularly efteemed and favored by two Popes, the only ones who filled the chair of St. Peter in his time, and as great men as ever fat there fince that apoftle, if, at leaft, he ever did; one, in fhort, who could have been a Leonardo, a Michael Angelo, a Titian, a Correggio, a Parruegiano, an Annibal, a Rubens, or any other whom he pleafed, but none of them could ever have been a Raphael."

He is allowed, writes Pilkinton, "to have diffufed throughout all his works, more grace, truth, and fublimity than any other painter, who has appeared before or fince."

"It was one of the remarkable properties of Raphael's genius," fays De Quincey, "that in the execution of his works he always expreffed, in a prominent manner, the greateft and moft elevated feature of his fubject, without, in any degree, fcorning the minuteft details. Lanzi has obferved, on this point, that the finifh he has given to his heads is fuch, that you can almoft count every particular hair."

"Michael Angelo," remarks Hazlitt, ("Criticifms on Art,") "was painter, fculptor, architect. Raphael was only a painter, but in that one art he feemed to pour out all the treafures and various excellences of nature, grandeur and fcope of defign, exquifite finifhing, force, grace, delicacy, the ftrength of man, the foftnefs of woman, the playfulnefs of infancy, thought, feeling,

invention, imitation, labor, eafe, and every quality that
can diftinguifh a picture, except color."

The grace and naturalnefs of the pictures of this
mafter are everywhere borne witnefs to.

"All great actions are fimple," fays Emerfon,
"and all great pictures are. The Transfiguration,* by
Raphael, is an eminent example of this peculiar merit.
A calm, benignant beauty fhines all over this picture,
and goes directly to the heart. It feems almoft to call
you by name. The fweet and fublime face of Jefus is
beyond praife, yet how it difappoints all florid expecta-
tion. This familiar, fimple, home-fpeaking countenance is as if one fhould meet a friend."

It would be eafy to heap up teftimony of a fimilar
character, to any extent, but the tafk is unneceffary.
Mrs. Bray, the accomplifhed biographer of Stothard,
the artift—well known by his numerous drawings,
efpecially the inimitable "Pilgrimage to Canterbury,"
and "Flitch of Bacon," fays of her fubject:—"There
can be no doubt that Stothard's youthful ftudy of Ra-

---

* This wonderful picture has been reproduced in Rome, in mofaics, at a coft of 12,000
crowns, and the labor of nine years; ten men working at it. The fmalts, of which thefe
mofaic pictures are formed, are a fpecies of opaque vitrified glafs, partaking of the mixed
nature of ftone and glafs. Of thefe, no lefs than feventeen hundred different fhades are in
ufe; they are manufactured in Rome, in the form of long flender rods, like wires, of different
degrees of thicknefs.

3

phael helped, not merely to form his tafte, but to de-
velope his own remarkable powers, and to make him
what he was. He had imbibed that grace and myftery
of painting which is fo tranfcendently beautiful in the
pictures of the Italian mafters. The Holy Families of
the Englifh painters are human beings; with the Italians they are only human forms, having, however, in-
fufed into them fomething of a fuperhuman fpirit."

It is known, fays De Quincey, that Raphael had a
fpecial devotion for the Virgin; this is attefted, in a
meafure, by his founding, in her honor, a chapel in the
church of Santa Maria della Rotunda, where, as we
have fhown, his afhes now repofe. But nothing, con-
tinues De Quincey, fo clearly manifefts in him the
various feelings of a piety, fometimes fimple and affectionate, fometimes full of grandeur and elevation, than
that diverfity of afpects under which his pencil, always
noble, though the fubject of the compofition be fimple,
always amiable and graceful though it be fublime, has
delighted in fetting forth, according to the taftes or
deftination for which they were intended, the image of
the Virgin——here, as the modeft inhabitant of Bethle-
hem——there, as the queen of the angels.

"His Madonnas," remarks Vafari, "difplay all
that the higheft idea of beauty could imagine in the

reprefentation of a youthful virgin: modefty in her eyes, on her forehead honor, in the line of the nofe grace, in the mouth virtue."

Hillard, in his criticifm upon the pictures in the Tribune, at Florence, bears witnefs to Raphael's match-lefs fkill in impreffing upon his productions that unde-finable grace and majefty which diftinguifh his works from all others. "Maternal love, purity of feeling, fweetnefs, refinement, and a certain foft ideal happinefs breathe from his canvas like odor from a flower. No painter addreffes fo wide a circle of fympathies as he ; no one fpeaks a language fo intelligible to the common apprehenfion."

The fecret of this wonderful fuccefs, fo far at leaft as the fecrets of genius can be penetrated, would feem to be, that Raphael never copied, but painted always, as indeed he has himfelf declared, from an idea in his own mind : while the Madonnas of moft other artifts were portraits. Andrea del Sarto, Rubens, and Albano, painted their wives ; Allori and Vandyck their mis-treffes ; Domenichino his daughter.

On this point, Kugler obferves : " Like all other artifts, Raphael is always greateft when, undifturbed by foreign influence, he follows the free and original im-pulfe of his own mind. His peculiar element was

grace and beauty of form, in as far as thefe are the expreffion of high moral purity.   Hence, notwithftanding
the grand works in which he was employed by the
Popes, his peculiar powers are moft fully developed in
the Madonnas and Holy Families, of which he has left
fo great a number.   In his youth, he feems to have
been fondeft of this clafs of fubjects.   They are conceived with a graceful freedom, fo delicately controlled,
that it appears always guided by the fineft feeling for
the laws of art.   They place before us thofe deareft
relations of life which form the foundation of morality,
the clofeft ties of family love; yet they feem to breathe
a feeling ftill higher and holier.   Mary is not only the
affectionate mother; fhe appears, at the fame time,
with an expreffion of almoft virgin timidity, and yet as
the bleffed one of whom the Lord was born.   The infant Chrift is not only the cheerful, innocent child, but
a prophetic ferioufnefs refts on his features which tells
of his future deftiny.

In any comparative eftimate of Raphael's powers or
performances, the fhortnefs of his life muft not be over-
looked.   It is fomewhat remarkable that of feventy-
feven artifts of renown, from Cimabue, born in 1240,
to Turner, who died in 1852, all but two—Paul Potter,
who died at the age of 27, and Giorgione, who lived

but 34 years——exceed Raphael in the length of their several careers.   The average life of the seventy-seven was 68 years, 8 months.

The eminent historical painter, Opie, concludes a lecture at the Royal Institution thus:——"The history of no man's life affords a more encouraging and instructive example than that of RAPHAEL.   The path by which he ascended to eminence is open, and the steps visible to all.   He began with apparently no very uncommon fund of ability, but, sensible of his deficiencies, he lost no opportunity of repairing them.   He studied all the artists of his own and former times, and penetrated all their mysteries, mastered their peculiarities, and grafted all their excellencies on his own stock."

Hymn to the Virgin.

EDGAR A. POE.

A T morn, at noon, at twilight dim,
Maria, thou haſt heard my hymn:
In joy and woe, in good and ill,
Mother of God, be with me ſtill !
When the hours flew brightly by,
And not a cloud obſcured the ſky,
My ſoul, leſt it ſhould truant be,
Thy grace did guide to thine and thee.
Now, when ſtorms of fate o'ercaſt
Darkly my preſent and my paſt,
Let my future radiance ſhine
With ſweet hopes of thee and thine.

Vierge au Berceau.

THIS charming conception was prefented by Raphael to Adrian Gouffier, Cardinal de Boiffy, whom Leo X. fent legate into France. It is painted on copper; is one foot three inches high, and eleven and one-half inches wide. After being preferved for a feries of years in the family of the recipient, it came into the cabinet of the Duke de Rouanez, and was purchafed by Louis XIV. of the Abbe de Brienne; and it now beautifies the walls of the Louvre. On the right, the infant Jefus, flanding, leaning on the Virgin, his feet refting on his cradle, takes in his hands the head of the young Saint John, whom Saint Elizabeth, kneeling, is prefenting to him. Behind the figures are trees and part of a wall in ruins. On the right and left, a beautiful landfcape.

De Quincey remarks of this picture, that in it "there is great vigor of tone and moſt careful handling. The genius of Raphael ſhines forth from every figure. The infant Jeſus is imbued with a grace and beauty truly divine. The landſcape is ſmiling and brilliant."

Invocation to the Virgin.

CHAUCER.

MODERNIZED BY WORDSWORTH.

MOTHER Maid! O Maid and Mother free!
  O bush unburnt, burning in Moses' sight!
That down didst ravish from the Deity,
  Through humblenefs, the Spirit that did alight
  Upon thy heart, whence, through that glory's
    might,
Conceived was the Father's fapience,
Help me to tell it in thy reverence!

Lady, thy goodnefs, thy magnificence,
  Thy virtue, and thy great humility,
Surpafs all fcience and all utterance;
  For, fometimes, Lady! ere men pray to thee,
  Thou go'ft before in thy benignity,

4

The light to us vouchfafing of thy prayer,
To be our guide unto thy Son fo dear.

My knowledge is fo weak, O blissful Queen,
    To tell abroad thy mighty worthinefs,
That I the weight of it may not fuftain,
    But as a child of twelve months old, or lefs,
    That laboreth his language to exprefs,
Even fo fare I; and therefore, I thee pray,
Guide thou my fong, which I of thee fhall fay.

La Madonna dell Pesce.

(SEE FRONTISPIECE.)

THE Madonna of the Fiſh was painted on panel, between 1513 and 1515, for the church of San Domenico, at Naples, and placed in that chapel wherein is the crucifix which ſpoke to St. Thomas Aquinas. By the chances of events, the picture was tranſported from Naples to Spain, from Spain to Paris, where it was transferred* from panel to canvas,

---

* The tranſfer of a painting from panel to canvas ſeems ſo impoſſible an operation, and the proceſs is ſo ingenious and intereſting, that it may not be amiſs to record here the deſcription of it, given by the members of the National Inſtitute, Paris, by whom it was performed upon another of Raphael's pictures:—

"It was neceſſary, as a previous ſtep, to render the ſurface of the panel, on which the picture was painted, perfectly plane. To this end, a gauze having been paſted over the painting, the picture was turned on its face. There was then formed in the ſubſtance of the wood a number of ſmall channels, at certain diſtances from each other, and extending from the upper extremity of the arch, to where the panel preſented a truer ſurface. He introduced into theſe channels ſmall wooden wedges, and afterwards covered the whole ſurface with wet cloths, which he took care to renew from time to time.

and again returned to Spain, where it now repofes in
the Gallery of the Efcurial in Madrid.   It reprefents
the Madonna and child upon a throne; on one fide,
and kneeling on a ftep of the throne, is St. Jerome,

"The action of thefe wedges, expanding by the humidity, obliged the panel to reaffume
its original form, the two parts of the crack before mentioned were brought together; and the
artift, having introduced a ftrong glue to re-unite them, applied crofs bars of oak, for the purpofe
of retaining the picture, during its drying, in the form which it had taken.

"The deficcation was performed very flowly; a fecond gauze was applied over the former,
and upon that two fucceffive layers of fpongy paper.   This preparation, which is called the
cartonnage, being dry, the picture was again inverted upon a table, to which it was firmly fixed
down, and they afterwards proceeded to the feparation of the wood on which the picture had
been painted.

"The firft operation was performed by means of two faws, the one of which worked per-
pendicularly, and the other horizontally.   The work of the faws being finifhed, the wood was
found to be reduced to one-tenth of an inch in thicknefs.   The artift afterwards made ufe of
a plane, of a convex form, in the direction of its breadth : this was applied obliquely upon the
wood, fo as to take off very fmall fhavings, and to avoid raifing the grain of the wood, which
was reduced by this means to ·002 of an inch thick.

"He took afterwards a flat-toothed plane, of which the effect is nearly fimilar to that of a
rafp, which takes off the wood in form of a duft or powder: it was reduced by this tool to a
thicknefs not exceeding that of an ordinary fheet of paper.

"In this ftate, the wood having been repeatedly wetted with fair water, in fmall compart-
ments, was carefully detached by the artift with the rounded point of a knife blade.   The
citizen Hacquin having then taken away the whole of the priming on which the picture had
been painted, and efpecially the varnifhes, which fome former reparations had made neceffary,
laid open the very fketch itfelf of Raffaello.

"In order to give fome degree of fupplenefs to the painting, fo much hardened by time, it
was rubbed with cotton dipped in oil, and wiped with old muffin; after which, a coating of
white lead, ground with oil, was fubftituted for the former priming, and laid on with a foft
brufh.

"After three months drying, a gauze was pafted on to the oil-priming, and over that a fine
cloth.   This being again dried, the picture was detached from the table, and again turned, for
the purpofe of taking off the cartonnage by means of water; which operation being finifhed,
they proceeded to take away certain inequalities of the furface, which had arifen from its un-

reading from a book. On the other fide, the young
Tobit (Tobias), bearing a fifh in one hand, is prefented
by the guardian angel Raphael.
"Tobias with his fifh," says Mrs. Jamefon, "was
an early type of baptifm." "The angel Raphael leading
Tobias, always expreffes protection, and efpecially protection to the young." Bonnemaifon, a learned com-
mentator, has pretended that the object of this picture
was to fignify the acknowledged canonicalnefs of the
Book of Tobit, and the verfion of it, made by St.
Jerome ; the child Jefus, by the reception he feems to
give to the young Tobit—expreffing the approbation
of the book by the Church. This is pronounced by

---

equal fhrinking during the former operations. To this end, the artift applied fucceffively to
thefe inequalities a thin pafte of wheaten flour, over which a ftrong paper being laid, he paffed
over it a heated iron, which produced the defired effect; but it was not until the moft careful
trial had been made of the due heat of the iron, that it was allowed to approach the picture.

"We have thus feen, that having fixed the picture, freed from every extraneous matter,
upon an oil priming, and having given a true form to its furface, it yet remained to apply this
chef-d'œuvre of art firmly upon a new ground. To this end, it was neceffary to paper it
afrefh, and to take away the gauze, which had been provifionally laid upon the priming, to add
a new coat of white lead and oil, and to apply upon that a very foft gauze, over which was
again laid a cloth, woven all of one piece, and impregnated on the exterior furface with a
refinous mixture, which ferved to fix it upon a fimilar cloth ftretched upon the frame. This
laft operation required the utmoft care, in applying to the prepared cloth the body of the painting, freed again from its cartonnage, in avoiding the injuries which might arife from too great
or unequal an extenfion, and, at the fame time, in obliging every part of its vaft extent to
adhere equally to the cloth ftretched upon the frame.

"Thus was this valuable picture incorporated with a bafe more durable even than its
former one, and guarded againft thofe accidents which had before produced its decay."

De Quincey, "one of Raphael's moft pleafing compofitions—one of thofe which appear to have been
moft completely the work of his own hand.   Its tone
is everywhere clear.   It has all the purity, all the fimplicity of the firft age; and, at the fame time, all the
firmnefs, all the breadth of ftyle, the fruit of mature
talent.   Nothing can be more true than the head of
Saint Jerome; nothing more expreffive than that of the
angel Raphael; nothing more fimple than the pofition,
or more innocent than the countenance of the young
Tobit; and never did the painter conceive any thing
more noble and more modeft, any thing grander and
more graceful, than the figure of the Virgin."

Wilkie fays, "the head and neck of the angel may
may be confidered to realize the beau-ideal of the fuppofed art of the Greeks."

Kugler confiders the picture as uniting "the fublime and abftract character of facred beings with the
individuality of nature in the happieft manner . . . all
the figures are graceful and dignified, and all combine
in beautiful harmony, and leave a refined impreffion on
the feelings of the fpectator."

Oh, Virgin Mother!

TRANSLATED FROM DANTE, BY CARY.

OH, Virgin-Mother, daughter of thy Son!
Created beings all in lowlinefs
Surpaffing, as in height above them all;
Term by the eternal counfel preordained;
Ennobler of thy nature, fo advanced
In thee, that its great Maker did not fcorn
To make himfelf his own creation;
For in thy womb, rekindling fhone the love
Revealed, whofe genial influence makes now
This flower to germin in eternal peace:
Here thou, to us, of charity and love
Art as the noonday torch, and art beneath,
To mortal men, of hope a living fpring.
So mighty art thou, Lady, and fo great,
That he who grace defireth, and comes not

To thee for aidance, fain would have defire
Fly without wings.   Not only him who afks,
Thy bounty fuccors; but doth freely oft
Forerun the afking.   Whatfoe'er may be
Of excellence in creature, pity mild,
Relenting mercy, large munificence,
Are all combined in thee !

La Vierge au Voile.

SEVERAL copies of this pleafing picture, or more properly, repetitions of the fame idea——the Sleeping Saviour, from whom the Holy Mother gently removes the covering——exift in the galleries of Europe. The one here reprefented is from the original in the Louvre, and is known alfo, as "La Vierge au Diademe," from the diadem with which the Virgin is crowned. In the eftimation of the editors of the great work, the "Mufee Francais," "this painting merits peculiar diftinction among the many Raphael executed on the fame fubject, from the beautiful fentiment it expreffes, and by the charm of the compofition. He has depicted the fweet fenfation of a tender mother when fhe contemplates her child funk in a deep and tranquil fleep. He has placed the Virgin crouched befide her infant, in the Eaftern manner, raifing foftly the veil that covers him, that he may be feen by St.

5

John. The background of the picture reprefents the ruins of a temple in the neighborhood of the town of Saccheti, near St. Peter. The picture belonged formerly to M. de la Vrilliere, and afterwards paffed into the cabinet of the Prince de Cavignan, and at his death was purchafed by Louis XIV."

The original is two feet two and three-fourths inches, by one foot feven and one-half inches, and, according to Kugler, has been much injured, like fo many others at the Louvre.

Mrs. Jamefon confiders the picture replete with grace and expreffion.

Can we better conclude, than by an extract from Mrs. Browning's Addrefs of the Virgin Mary to the Child Jefus?—"Sleep, fleep, mine Holy One"—

"Perchance this fleep that fhutteth out the dreary
Earth founds and motions, opens on thy foul
        High dreams on fire with God;
High fongs that make the pathways where they roll
More bright than ftars do theirs; and vifions new
Of thine eternal nature's old abode.
        Suffer this mother's kifs,
        Beft thing that earthly is,
To glide the mufic and the glory through,
To narrow in thy dream the broad upliftings
        Of any feraph's wing.
Thus, noifelefs, thus! Sleep, fleep, my dreaming One."

The Worship of the Madonna.

MRS. JAMESON.

OF the pictures in our galleries, public or private—
of the architectural adornments of thofe majeftic
edifices which fprung up in the middle ages
(where they have not been defpoiled or defe-
crated by a zeal as fervent as that which reared them),
the largeft and moft beautiful portion have reference to
the Madonna—her character, her perfon, her hiftory.
It was a theme which never tired her votaries—
whether, as in the hands of the great and fincere
artifts, it became one of the nobleft and lovelieft, or, as
in the hands of the fuperficial, unbelieving, time-
ferving artifts, one of the moft degraded. * * * It is
not my intention to enter here on that difputed point,
the origin of the worfhip of the Madonna. * * * That

the veneration paid to Mary in the early Church was a very natural feeling in thofe who advocated the divinity of her Son, would be granted, I fuppofe, by all but the moft bigoted reformers; that it led to unwife and wild extremes, confounding the creature with the Creator, would be admitted, I fuppofe, by all but the moft bigoted Roman Catholics.

How it extended from the Eaft over the nations of the Weft, how it grew and fpread, may be read in ecclefiaftical hiftories. Everywhere it feems to have found in the human heart fome deep fympathy— deeper far than mere theological doctrine could reach —ready to accept it; and in every land the ground prepared for it in fome already dominant idea of a Mother-Goddefs, chafte, beautiful, and benign. * * *

It is curious to obferve, as the worfhip of the Virgin-Mother expanded and gathered to itfelf the relics of many an ancient faith, how the new and the old ele-ments, fome of them apparently the moft heterogene-ous, became amalgamated, and were combined into the early forms of art;—how the Madonna, when fhe affumed the characterftics of the great Diana of Ephe-fus, at once the type of Fertility, and the Goddefs of Chaftity, became, as the imperfonation of motherhood, all beauty, bounty, and gracioufnefs; and at the fame time, by virtue of her perpetual virginity, the patroneſs of ſingle and aſcetic life—the example and the excuſe for many of the wildeſt of the early monkiſh theories. * * * The firſt hiſtorical mention of a direct worſhip paid to the Virgin Mary, occurs in a paſſage in the works of St Epiphanius, who died in 403. The very firſt inſtance which occurs in written hiſtory of an invocation to Mary, is in the life of St. Juſtina, as related by Gregory Nazianzen. To the ſame period—the fourth century—we refer the moſt ancient repreſentations of the Virgin in art. The earlieſt figures extant are thoſe on the Chriſtian ſarcophagi; but neither in the early ſculpture, nor in the moſaics of S. Maria Maggiore, do we find any figure of the Virgin ſtanding alone; ſhe forms a part of the group of the Nativity or the Adoration of the Magi. There is no attempt at individuality or portraiture. St. Auguſtine ſays expreſſly, that there exiſted, in his time, no authentic portrait of the Virgin.

Holy Family.

FROM THE GERMAN OF GOETHE.

CHILD of beauty rare—
O mother chaste and fair—
How happy seem they both, so far
    beyond compare!

She, in her infant blest,
And he in conscious rest,
Nestling within the soft warm cradle
    of her breast!

What joy that sight might bear
To him who sees them there,
If with a pure and guilt untroubled eye,
He looked upon the twain, like Joseph
    standing by.

Madonna della Seggiola.

THIS celebrated picture—entitled alfo, "La Vierge a la Chaife"—is, without exception, the beft known of Raphael's Madonnas, and that from which the greateft number of copies have been taken. It is, therefore, inconteftably the favorite with the public, if not with artifts and amateurs.

This has been varioufly accounted for. A modern writer on Art, remarks of the Virgin-Mother (whofe fitting pofition, it may be obferved, gives the picture its diftinctive appellation), "Her form, her features, an indefcribable fweetnefs of expreffion, the maternal tendernefs beaming from her foft hazel eye, the modeft and pious confcioufnefs of being the mother of a God, the pofition of the child's cheek to her own, expreffive at once of both dignity and fondnefs of affection, the

propriety of coftume, the coloring, the finifh—all, all
are divine."

The Editors of the famous " Mufee Francais "
difcourfe thus :—" All thefe pictures of Raphael are
conceived with judgment, compofed with grace, drawn
with precifion, and painted with the utmoft perfection
of art. Whence comes it, then, that this, more than
any other, poffeffes an inconceivable charm, but from
the countenance of the Virgin, whofe features are
more uniformly fine, whofe eyes have greater vivacity,
whofe whole expreffion is more ftriking and gracious,
than diftinguifh any other compofition on the fame
fubject, which are more generally remarked for fimplicity of character.

" The contouring, likewife, exhibits extraordinary
purity, correctnefs, and beauty. It is remarked that
the paint itfelf is fuperior to that employed by Raphael
in any other production."

De Quincey confiders this, in " coloring and grace
of attitude and arrangement, one of Raphael's moft
agreeable productions. The manner in which the
child and mother are grouped, and in which the head
of the latter is turned back, the elegance and grace of
the enfemble, have fingularly captivated the tafte of
thofe who are lefs fenfible to the religious keeping of

the fubject, than to the general impreffion of a graceful
effect upon the fenfes."

The accomplifhed author of the "Six Months in
Italy" regards it as a work of great fweetnefs, purity,
and tendernefs, but not reprefenting all the power of
the artift's genius. "Its chief charm, and the fecret of
its world-wide popularity, is its happy blending of the
divine and the human elements. Some painters treat
this fubject in fuch a way that the fpectator fees only a
mortal mother careffing her child; while, by others,
the only ideas awakened are thofe of the Virgin and
the Redeemer. But heaven and earth meet on Raphael's canvas : the purity of heaven and the tendernefs
of earth. The round, infantile forms, the fond, clafp-
ing arms, the fweetnefs and the grace belong to the
world that is around us ; but the faces——efpecially that
of the infant Saviour, in whofe eyes there is a myfterious
depth of expreffion, which no engraving has ever fully
caught——are touched with light from heaven, and fug-
geft fomething to worfhip as well as to love."

Mrs. Jamefon, in her "Diary of an Ennuyee,"
records of this Madonna :——"The prevailing expreffion
is a ferious and penfive tendernefs; her eyes are turned
from her infant, but fhe clafps him to her bofom, as if
it were not neceffary to fee him, to feel him in her heart."

6

And laftly, Kugler defcribes her as " a beautiful and blooming woman, looking out of the picture in the tranquil enjoyment of maternal love; the Child is full and ftrong in form, has a ferious, ingenuous and grand expreffion. The coloring is uncommonly warm and beautiful."

The original is circular in form, two feet four inches in diameter. It was painted about 1516, and formed part of the Florentine Gallery from 1539 till a later period, when it was transferred to the Pitti Palace.

It has been valued at 150,000 francs.

Mary!

SHELLEY.

SERAPH of Heaven! too gentle to be human,
Veiling beneath that radiant form of woman
All that is infupportable in thee
Of light, and love, and immortality!
Sweet benediction in the eternal curfe!
Veiled Glory of this lamplefs univerfe!
Thou Moon beyond the clouds! Thou living
     Form
Among the Dead! Thou Star above the ftorm!
Thou Wonder, and thou Beauty, and thou
     Terror!
Thou Harmony of Nature's art! Thou Mirror
In whom, as in the fplendor of the Sun,
All fhapes look glorious which thou gazeft on!

See where she stands! a mortal shape endued
With love, and life, and light, and deity;
The motion which may change but cannot die;
An image of some bright eternity;
A shadow of some golden dream; a splendor
Leaving the third sphere pilotless.

La Vierge aux Palmiers.

ON the firft vifit of Raphael to Florence, he was welcomed with warm hofpitality by Zaddeo Taddei, a great admirer of genius. Raphael, that he might not be furpaffed in generofity and courtefy, painted, probably between 1506 and 1508, two pictures for his kind entertainer, wherein there are traces of his firft manner, derived from Pietro, and alfo of that much better one which he acquired by ftudy. Thefe were both pictures of the Madonna, and after the deceafe of Zaddeo's immediate heirs, were difperfed, and only traced within a few years. One is in the Gallery of the Belvidere at Vienna; the other, reprefenting the entire Holy Family repofing under a palm tree, is in the Bridgewater Gallery, in the poffeffion of the Earl of Ellefmere, London. It was formerly in the Orleans Collection, having been purchafed for 1,000 pounds.

It is circular in form, three feet nine inches in diameter; was originally painted on panel, but since transferred to canvas.

Not wholly inappropriate, in this connection, are Mrs. Hemans' fine lines on the " Repose of the Holy Family, during the Flight into Egypt :"—

> " Under a palm tree, by the green old Nile,
>   Lulled on his mother's breast, the fair child lies,
> With dove-like breathings, and a tender smile
>   Brooding above the slumber of His eyes;
> While, through the stillness of the burning skies,
>   Lo! the dread works of Egypt's buried kings,
> Temple and pyramid, beyond Him rise,
>   Regal and still as everlasting things.
> Vain pomps! from Him, with that pure flowery cheek,
>   Soft shadowed by His mother's drooping head,
> A new born spirit, mighty and yet meek,
>   O'er the whole world like vernal air shall spread,
> And bid all earthly grandeurs cast the crown,
>   Before the suffering and the lowly down."

Raphael and Fornarina.

BY L. E. LANDON.

[RAPHAEL was essentially the painter of beauty. Of the devotion with which he sought its inspiration, in its presence, a remarkable instance is recorded. He either could not, or would not, paint without the presence of his lovely mistress, LA FORNARINA.]

AH! not for him the dull and measured eye,
Which colors nothing in the common sky,
Which sees but night upon the starry cope,
And animates with no mysterious hope.
Which looks upon a quiet face, nor dreams
If it be ever tranquil as it seems;
Which reads no histories in a parting look,
Nor on the cheek, which is the heart's own book,
Whereon it writes in rosy characters
Whate'er emotion in its silence stirs.

Such are the common people of the foul,
Of whom the ftars write not in their bright fcroll.
Thefe, when the funfhine on the noontide makes
Golden confufion in the foreft brakes,
See no fweet fhadows gliding o'er the grafs,
Which feem to fill with wild flowers as they pafs;
Thefe, from the twilight mufic of the fount
Afk not its fecret and its fweet account;
Thefe never feek to read the chronicle
Which hides within the hyacinth's dimlit bell:
They know not of the poetry which lies
Upon the fummer rofe's languid eyes;
They have not fpiritual vifitings elyfian,
They dream no dreamings, and they fee no vifion.
　　The young Italian was not of the clay,
That doth to duft one long allegiance pay.
No; he was tempered with that finer flame,
Which ancient fables fay from heaven came;
The funfhine of the foul, which fills the earth
With beauty borrowed from its place of birth.
Hence has his lute its fong, the fcroll its line;
Hence ftands the ftatue glorious in its fhrine;
Hence the fair picture, kings are fain to win,
The mind's creations from the world within.

*　　*　　*　　*　　*　　*

Not without me !—alone, thy hand
   Forgot its art awhile ;
Thy pencil loſt its high command,
   Uncheriſhed by my ſmile.
It was too dull a taſk for thee
   To paint remembered rays ;
Thou, who were wont to gaze on me,
   And color from that gaze.

I know that I am very fair,
   I would I were divine,
To realize the ſhapes that ſhare
   Thoſe midnight hours of thine.
Thou ſometimes telleſt me, how in ſleep
   What lovely phantoms ſeem ;
I hear thee name them, and I weep
   Too jealous of a dream.

But thou did'ſt pine for me, my love,
   Aſide thy colors thrown ;
'Twas ſad to raiſe thine eyes above,
   Unanſwered by mine own ;
Thou who art wont to lift thoſe eyes,
   And gather from my face
The warmth of life's impaſſioned dyes,
   Its color and its grace.

Ah! let me linger at thy fide,
 And fing fome fweet old fong,
That tells of hearts as true and tried,
 As to ourfelves belong.
The love whofe light thy colors give,
 Is kindled at the heart;
And who fhall bid its influence live,
 My Raphael, if we part?

La Vierge a l'Oiseau.

KNOWN ALSO AS THE

MADONNA DEL CARDELLINO.

HILE Raphael was in Florence, for the firſt time, he formed a cloſe friendſhip with Lorenzo Naſi, and the latter having taken a wife at that time, Raphael, ſays Vaſari, painted a picture for him, wherein he repreſented Our Lady with the Infant Chriſt, to whom St. John, alſo a child, is joyouſly offering a bird, which is cauſing infinite delight and gladneſs to both children. In the attitude of each there is a child-like ſimplicity of the utmoſt lovelineſs; they are, be-ſides, ſo admirably colored, and finiſhed with ſo much care, that they ſeem more like living beings than paint-ings. Equally good is the figure of the Madonna: it has an air of ſingular grace and even divinity, while all the reft of the work—the foreground, the furrounding landfcape, and every other particular, are extremely beautiful. This picture was held in the higheft eftimation by Lorenzo Nafi fo long as he lived, not only becaufe it was a memorial of Raphael, who had been fo much his friend, but on account of the dignity and excellence of the whole compofition; but on the 9th of Auguft, 1548, the work was nearly deftroyed by the finking down of the hill of San Giorgio, when the manfion of Lorenzo was overwhelmed by the fallen maffes. The fragments of the picture were found among the ruins of the houfe, and put together in the beft manner that he could contrive by Batifta, a fon of Lorenzo, who was a great lover of art! The picture now adorns the Tribune of the Florentine Gallery, though this has been regarded by fome as a duplicate, or perhaps a copy, of the original work prefented to Nafi.

Hillard, in that charming record of his " Six Months in Italy," alluding to this picture in connection with a " St. John in the Defert," alfo by Raphael— remarks, " Thefe two pictures are not penetrated with that maturity and vigor which Raphael's genius fubfequently attained, but they are full of thofe winning and engaging qualities which belonged to it in every ftage of its development."

Mrs. Jamefon regards this work as perhaps the moſt perfeĉt example of the claſs of Madonnas to which it belongs—the group of three—which could be cited from the whole range of art: and Kugler ſays, "The form and countenance of the Madonna are of the pureſt beauty; her whole ſoul ſeems to breathe holineſs and peace. John alſo is extremely ſweet; but the figure of the infant Chriſt does not fulfil the artiſt's intention, which appears to have been to repreſent the ſerioufneſs and dignity of a Divine being in a childlike form."

Zappi's Sonnet

ON THE PORTRAIT OF RAPHAEL BY HIMSELF.

TRANSLATED FROM THE ITALIAN, BY

GEO. W. BETHUNE.

AND this is Raffaelle! There in that one face,
　　So sadly sweet, sought nature to portray
　　His own high dreams of noblenefs and grace,
　　The all of genius that she could convey
In features visible. He alone could trace
　The great Idea; nor could he essay
Upon the eternal canvas thus to place,
　Secure in beauty far beyond decay,
Another form so glorious as his own.
　E'en eager Death held in suspense his dart:
"How shall the painter from his work be known?"
　He asks, "that I may strike him to the heart?"
"Fruitless thy rage," the great soul gives reply,
　"Nor image, nor its author, e'er shall die!"

Raphael's Genius.

ILTON has been compared to Raphael. He is much more like Michael Angelo. Michael Angelo is the painter of the Old Teftament, Raphael of the New. Now Milton, as Wordsworth has faid of him, was a Hebrew in foul. He was grand, fevere, auftere. He loved to deal with the primeval, elementary forms, both of inanimate nature and of human, before the manifold, ever-multiplying combinations of thought and feeling had fhaped themfelves into the multifarious complexities of human character. * * * *

Where to find a parallel for Raphael in the modern world, I know not. Sophocles, among poets, moft refembles him. In knowledge of the diverfities of human character, he comes nearer than any other

painter to him, who is unapproached and unapproach-
able, Shakſpeare; and yet two worlds, that of Humor,
and that of Paſſion, ſeparate them.   In exquiſiteneſs
of art, Goethe might be compared to him.   But
neither he nor Shakſpeare has Raphael's deep Chriſtian
feeling.   And then there is ſuch a peculiar glow and
bluſh of beauty in his works: whitherſoever he comes,
he ſheds beauty from his wings.

Why did he die ſo early?   Becauſe morning cannot
laſt till noon, nor ſpring through ſummer.   Early, too,
as it was, he had lived through two ſtages of his art,
and had carried both to their higheſt perfection.

This rapid progreſſiveneſs of mind he alſo had in
common with Shakſpeare and Goethe, and with few
others.

Sainte Famille dite La Perle.

FTER the death of Charles I. of England, a sale was ordered of his collection of Works of Art, valued at £49,903 2s. 6d. The difperfion took place in 1650 and 1653, attracting vaft numbers of agents from foreign princes, and amateurs from all parts of Europe. The total proceeds of the fale, including the embroideries, jewels, etc., was £118,080 10s. 2d.; the feven Cartoons being purchafed for the Britifh nation for £300.

The purchafes of the Spanifh ambaffador, Don Alonzo de Cardenas, were fo great, that eighteen mules were required to convey them from Corunna to Madrid. Among them was the large Holy Family by Raphael, from the Mantua Collection, for which he gave £2,000. Philip IV. is faid to have exclaimed on feeing it, " This is my Pearl;" hence the picture has

8

been fo defignated by lovers of art. " And he was a
good judge," fays the editor of Murray's Hand-Book
of Spain, " for never was the ferious gentlenefs of the
bleffed Virgin-Mother, her beauty of form, her purity
of foul, better portrayed."

Says Kugler, it is "the moft important, and, in
compofition, unqueftionably the fineft of Raphael's
Holy Families. The figures arranged in perfect harmony, form a beautiful group."

The Madonna is reprefented full life-fize, holding
with one hand the Infant Jefus, who is half feated on her
right knee, his left leg refting on the cradle, the other
hanging down. The little St. John, raifing with both
hands the fkirt of his fkin garb, is prefenting to the
Infant Jefus the fruits he has collected there. The
child, ere he takes them, turns fmilingly towards his
mother, as if to folicit her permiffion. Mary's left arm
refts on the fhoulder of St. Anne, who, kneeling, feems
abforbed in meditation. The background is occupied
on one fide with a landfcape, on the other with ruins,
clofe to which we fee St. Jofeph.

De Quincey remarks, "The coloring of this picture,
though fomewhat faded by the effects of time, has preferved great vigor, and a harmony, which, in fome of
its parts, need fear no comparifon with the works of

the Venetian fchool. The flefh tints of the Infant
Jefus are as brilliant, as the movement and outlines of
the figure are graceful and pure. In more than one
place of the picture, we detect corrections, or fecond
thoughts. We learn from thefe that the head of the
Virgin, now a three-quarters face, was at firft in profile.
The hair above the left temple has been raifed. We
alfo perceive feveral alterations in the outline of the
left hand of the Virgin, and of the left thigh of the
child."

Three others of Raphael's pictures are in the Efcurial Gallery : i. e., The Madonna del Spafimo, Virgen
del Pez (of the Fifh), and the Annunciation. Thefe
four gems came near falling into Englifh hands a few
years fince. It having been intimated that the government was difpofed to part with them, Lord Clarendon
offered, through the Spanifh minifter, the fum of
£80,000 for them. But profound fecrecy was a condition of the negotiation—whether with a view of
replacing the originals with copies, on the walls of the
Gallery, which a public, ignorant of the fale, fhould
accept as genuine, can only be furmifed,—and the
matter coming to the ears of the public, the Spanifh
government withdrew its confent to the fale.

Studies of Raphael.

HENRY WARD BEECHER.

WHEN I was in the Galleries of Oxford, I saw many of the defigns of Raphael and Michael Angelo. I looked upon them with reverence, and took up fuch of them as I was permitted to touch, as one would take up a love-token. It feemed to me, thefe fketches brought me nearer the great masters than their finifhed pictures could have done, becaufe therein I faw the mind's proceffes as they were firft born. They were the firft falient points of the infpiration.

MOZART and RAPHAEL! as long as the winds make the air give forth founds, and the fun paints the earth with colors, fo long fhall the world not let thefe names die.

Hymn to the Virgin.

FRANCES SARGENT OSGOOD.

MOTHER of the spirit child!
　　Of the guileless and the meek,
　　Mournful are thine eyes, but mild
　　　With a beauty from above;
　　　Pale but eloquent with love,
　　Thy youthful brow and cheek!
Thou, oh! thou hast known a parent's wasting grief!
A suppliant parent kneels, imploring thy relief!

　　By the pure and solemn joy
　　　Filling all thy maiden breast,
　　When the precious heaven-born boy,
　　　Glowing with celestial charms,
　　　Lay within those virgin arms
　　　A bright and wondrous guest!
Hear in mercy, hear the faltering voice of grief!
A suppliant mother kneels, imploring thy relief!

By thine anguiſh in that hour,
   Hour of woe and dread, when Death
Dared to ſtay the awful power,
    High, majeſtic yet benign;
    Dared to ſeal the truth divine
   Which dwelt upon his breath!
By thy hope, thy truſt, thy rapture, and thy grief,
Oh! ſainted Marie! ſend this breaking heart relief!

Madonna di Foligno,

LA VIERGE AU DONATAIRE.

HIS renowned Madonna belongs to the clafs of votive * pictures. Sigifmund Conti, of Foligno, a learned hiftorian, and private fecretary to Pope Julius II., having been in great danger from a meteor, or thunderbolt, vowed an offering to the Bleffed Virgin, to whom he attributed his fafety, and in fulfilment of his vow, induced Raphael to paint this precious

---

* Providential efcapes, victories, and fucceffes, were among the moft frequent occafions of what are called votive pictures. In thefe compofitions, the Madonna and Child are generally reprefented furrounded by faints, the latter being felected for various reafons, according to the tafte or devotion of the proprietor of the picture. The donor is frequently introduced kneeling, fometimes alone, fometimes with his family, and in many cafes a patron faint recommends the votaries. The ultimate interceffion of the Madonna is, however, diftinctly intimated by her appearing in the character of the "Mater Dei." When fhe is reprefented alone, her action is more directly that of a fuppliant.—Sir Chas. Eastlake.

picture, which he beftowed upon the Church of Ara Cœli in Rome. This was in 1511, when Raphael was in his twenty-eighth year. In the upper part of the picture is the Madonna with the Child, enthroned on the clouds in a glory, furrounded by angels. Underneath, on one fide, kneels the donor, raifing his folded hands to the Virgin; behind him ftands St. Jerome, who recommends him to her care. On the other fide is St. Francis, alfo kneeling and looking upward, while he points with one hand out of the picture to the people, for whom he entreats the protection of the Mother of Grace; behind him is John the Baptift, who points to the Madonna, while he looks at the fpectator as if inviting the latter to pay her homage.

In the centre of the picture and immediately beneath the Virgin, is an angel boy; his head raifed, while in his hands he holds a tablet, evidently intended for an infcription, though no trace appears thereon. The background is a beautiful landfcape. In the diftance is the city of Foligno, on which falls a meteor, an allufion to the circumftance which called forth the donation; above thefe, arches a rainbow—pledge of peace and fafety.

The Church of Ara Cœli, in which the picture was dedicated, belonged to the Francifcans, which ac-

counts for the introduction of St. Francis into the composition. The presence of the figure of St. Jerome is not so easily explained; but Mrs. Jamefon suggests the following hypothesis, which is at least ingenious, and not improbable: " The patron faint of the donor, St. Sigifmund, was a king and a warrior, and Conti might possibly think that it did not accord with his profession, as an humble ecclesiastic, to introduce him here. The most celebrated convent of the Jeronimites in Italy is that of St. Sigifmund near Cremona, placed under the special protection of St. Jerome, who is also, in a general sense, the patron of all ecclesiastics; hence perhaps he figures here as the protector of Sigifmund Conti."

Conti died in 1512, and in 1565 the picture was removed by his grandniece, Inora Anna Conti, to a convent called Le Conteffe, at Foligno, of which she was a nun. It was carried off to Paris by the French in 1792. At the restoration of the works of art in Italy, in 1815, it was placed in the VATICAN, of which it is now one of the most prized ornaments. This picture has received the highest encomiums for its spirit and execution, in its several parts and as a whole. It has been pronounced "one of Raphael's most remarkable examples for the expression of character," and as

9

"one of the moft vigorous in coloring and general execution."

Vafari, in commenting upon the individual figures, remarks of St. John, " We recognize him by his attenuated frame, the refult of penitence and long fafting ; his countenance, the mirror of his foul, announces that franknefs and abruptnefs of manners ufual with thofe who flee the world, and who, if ever they appear in it, manifeft themfelves the enemies of all diffimulation."

To which fentiment, De Quincey (the Biographer of Raphael and Michael Angelo) adds : " This is what Pliny calls pingere mores—an expreffion, the literal tranflation of which does not adequately reprefent its meaning, which fhould be conftrued—to paint the moral of each fubject."

Hazlitt fays, " I know not enough how to admire the innumerable heads of cherubs furrounding her, touched in with fuch care and delicacy, yet fo as fcarcely to be perceptible except on clofe infpection, nor that figure of the winged cherub below, offering the cafket, and with his round chubby face and limbs as full of rofy health and joy, as the cup is full of the juice of the purple vines."

Kugler alfo fpeaks of the angel with the tablet, as " of unfpeakable intenfity and exquifite beauty—

one of the moft marvellous figures that Raphael has
created:" and Vafari confiders it "not poffible to imagine any thing more graceful or more beautiful than
this child, whether as regards the head or the reft of
the perfon."   .

The original is reported to be one of the beft preferved of Raphael's pictures, many of which have fuf-
fered from the hand of time, and the yet more to be
dreaded hand of picture-cleaners and "reftorers."

Stanzas,

BY A BEAUTIFUL COPY OF THE MADONNA AND CHILD.

BY BERNARD BARTON.

I MAY not change the simple faith,
  In which from childhood I was bred;
Nor could I, without scorn or scathe,
  The living seek among the dead;
My soul has far too deeply fed
  On what no painting can express,
To bend the knee, or bow the head,
  To aught of pictured loveliness.

And yet, Madonna! when I gaze
  On charms unearthly, such as thine;
Or glances yet more reverent raise,
  Unto that infant, so Divine!

I marvel not that many a fhrine
   Hath been, and ftill is, reared to thee,
Where mingled feelings might combine
   To bow the head and bend the knee.

For who——that is of woman born,
   And hath that birthright underftood,
Mindful of being's early morn,
   Can e'er behold with thoughtlefs mood,
Moft pure and perfect womanhood?
   Woman——by angel once addreffed;
And by the wife, the great, the good
   Of every age accounted blefled!

Or who that feels the fpell——which Heaven
   Cafts round us in our infancy,
But more or lefs, hath homage given
   To childhood——half unconfcious why?
A yet more touching myftery
   Is in that feeling comprehended,
When thus is brought before the eye,
   Godhead with childhood ftrangely blended.

And hence I marvel not at all,
   That fpirits, needing outward aid,

Should feel and own the magic thrall
  In your meek lovelinefs difplayed :
And if the objects thus portrayed
  Brought comfort, hope, or joy to them,
Their error, let who will upbraid,
  I rather pity——than condemn.

For me, though not by hands of mine
  May fhrine or altar be upreared ;
In you, the human and Divine
  Have both fo beautiful appeared,
That each, in turn, hath been endeared,
  As in you feeling has explored
Woman——with holier love revered,
  And God——more gratefully adored.

La Vierge aux Candelabres.

THIS fomewhat fingular reprefentation of the Holy Mother and Child, is claffed by Kugler among thofe Madonnas which were, in a great meafure, the work of Raphael's fcholars, and only partially touched by the fingers of the great mafter himfelf.

If the face of the infant is one of the leaft pleafing of the feries, his pofition is eafy and natural, and the countenance of the Virgin has a large fhare of that ineffable fweetnefs of expreffion, combined with a calm thoughtfulnefs, which characterizes all the Madonnas of this artift.

The original painting formed a part of the collection of Lucien Bonaparte, Prince of Canino, at the Palazzo Lucano, Rome; and was fold in England, with other treafures of art, by the Duke of Lucca, in 1840, and is now in the poffeffion of Mr. Munro.

The Legend of Santarem.

BY CAROLINE SOUTHEY.

OME liften to a monkifh tale of old,
  Right Catholic, but puerile fome may deem,
Who all unworthy their high notice hold
  Aught but grave truth, or lofty learned theme;
Too wife for fimple fancies, fmiles and tears,
Dreams of our earlieft, pureft, happieft years.

Come—liften to my legend; for of them
  Surely thou art not: and to thee I'll tell
How on a time in holieft Santarem
  Strange accident miraculous befell
Two little ones; who to the facred fhrine
Came daily to be fchooled in things divine.

Twin fifters—orphan innocents were they :
  Moft pure, I ween, from all but the olden taint,
Which only Jefu's blood can wafh away :
  And holy as the life of holieft faint,
Was his, that good Dominican's, who fed
His mafter's lambs, with more than daily bread.

The children's cuftom, while that pious man
  Performed the various duties of his ftate
Within the fpacious church, as facriftan,
  Was on the altar fteps to fit and wait,
Neftling together ('twas a lovely fight !)
Like the young turtle-doves of Hebrew rite.

A fmall rich chapel was their fanctuary,
  While thus abiding ;—with adornment fair
Of curious carved work, wrought cunningly,
  In all quaint patterns, and devices rare :
And over them, above the altar, fmiled
From Mary-Mother's arms, the Holy Child.

Smiled on his infant guefts, as there below,
  On the fair altar fteps, thefe young ones fpread
  (Nor aught irreverent in fuch act I trow)
  Their fimple morning meal of fruit and bread.
  10

Such feaſt not ill beſeemed the ſacred dome—
Their father's houſe is the dear children's home.

At length it chanced, upon a certain day,
   When Frey Bernardo to the chapel came,
Where patiently was ever wont to ſtay
   His infant charge; with vehement acclaim
Both liſping creatures forth to meet him ran,
And each to tell the ſame ſtrange tale began.

" Father !" they cried, as hanging on his gown
   On either ſide, in each perplexed ear
They poured their eager tidings—" He came down—
   Menino Jeſu has been with us here !—
We aſked him to partake our fruit and bread;
And he came down—and ſat with us—and fed."

"Children ! my children ! know ye what ye ſay?"
   Bernardo haſtily replied—" But hold !—
Peace, Briolanja ! raſh art thou alway :
   Let Inez ſpeak." And little Inez told,
In her flow ſilvery ſpeech diſtinctly o'er,
The ſame ſtrange tidings he had heard before.

"Bleffed are ye, my children!" with devout
  And deep humility, the good man cried—
"Ye have been highly favored. Still to doubt
  Were grofs impiety and fceptic pride.
Ye have been highly favored. Children, dear!
Now your old mafter's loving counfel hear.

"Return to-morrow with the morning light,
  And as before, fpread out your fimple fare
On the fame table; and again invite
  Menino Jefu to defcend and fhare:
And if he come, fay—'Bid us, bleffed Lord!
We and our mafter to thy heavenly board.'

"Forget not, children of my foul! to plead
  For your old mafter: even for his fake
Who fed ye faithfully: and he will heed
  Your innocent lips; and I fhall fo partake
With his dear lambs. Beloved, with the fun
Return to-morrow. Then—His will be done."

"To-night! to-night! Menino Jefu faith
  We fhall fup with him, Father! we and thee,"
Cried out both happy children in a breath
  As the good father entered anxioufly

About the morrow's noon, that holy fhrine,
Now confecrate by fpecial grace divine.

"He bade us come alone; but then we faid
    We could not, without thee, our Mafter dear—
At that, he did not frown, but fhook his head
    Denyingly: Then ftraight with many a tear
We prayed fo fore, he could not but relent,
And fo he fmiled at laft, and gave confent."

"Now God be praifed!" the old man faid, and fell
    In prayer upon the marble floor ftraightway,
His face to earth: and fo, till vefper bell,
    Entranced in the fpirit's depths he lay;
Then rofe like one refrefhed with wine, and ftood,
Compofed among th' affembling Brotherhood.

The mafs was faid; the evening chant was o'er;
    Hufhed its long echoes through the lofty dome:
And now Bernardo knew the appointed hour
    That he had prayed for, of a truth was come.
Alone he lingered in the folemn pile,
Where darknefs gathered faft from aifle to aifle;

Except that through a diſtant door-way ſtreamed
  One ſlanting ſunbeam, gliding whereupon
Two angel ſpirits—(ſo in ſooth it ſeemed,
  That lovelieſt viſion)—hand in hand come on,
With noiſeleſs motion. " Father I we are here,"
Sweetly ſaluted the good Father's ear.

A hand he laid on each fair ſun-bright head,
  Rayed like a ſeraph's with effulgent light,
And—" Be ye bleſt, ye bleſſed ones," he ſaid,
  " Whom Jeſu bids to his own board to-night—
Lead on, ye choſen, to the appointed place
Lead your old maſter." So, with ſtedfaſt face,

He followed, where theſe young ones led the way
  To that ſmall chapel—like a golden clue
Streamed on before that long bright ſunſet ray,
  Till at the door it ſtopt.   Then paſſing through,
The maſter and the pupils, ſide by ſide,
Knelt down in prayer before the Crucified.

Tall tapers burnt before the holy ſhrine ;
  Chalice and paten on the altar ſtood,
Spread with fair damaſk.   Of the crimſon wine
  Partaking firſt alone ; the living food

Bernardo next with his dear children fhared—
Young lips, but well for heavenly food prepared.

And there we leave them.   Not for us to fee
    The feaft made ready, that firft act to crown;
Nor to perufe the folemn myftery
    Of the divine Menino's coming down
To lead away th' elect, expectant three,
With him that night at his own board to be.

Suffice it, that with him they furely were
    That night in Paradife; for thofe who came
Next to the chapel found them as in prayer,
    Still kneeling—ftiffened every lifelefs frame,
With hands and eyes upraifed as when they died,
Toward the image of the Crucified.

That mighty miracle fpread far and wide,
    And thoufands came the feaft of death to fee;
And all beholders, deeply edified,
    Returned to their own homes more thoughtfully,
Mufing thereon: with one great truth impreft,
That "to depart and be with Chrift is beft."

La Belle Jardiniere.

SECOND only to the Madonna della Seggiola, in popular eftimation, this fimple and beautiful picture has been reproduced in numerous engravings, and become one of the beft known of the feries. "The fweeteft cheerfulnefs, grace, and innocence," obferves Kugler, "breathe from this picture."

It was painted by Raphael at Florence, juft as he was leaving that city for Rome, about 1507, and fent to Siena. It was purchafed by Francis I., and now forms one of the attractions of the Louvre,—though it has unfortunately been much injured.

De Quincey defcribes it as "one of thofe fimple defigns, which more efpecially from the fize (fmall life) of the figures, we may place in the firft rank of thofe in which Raphael, before rifing to the ideal of his fubject, as he afterwards did, confined himfelf to the conceptions of pure fimplicity, innocence, and modeft grace, of which he found the models in the young village girls.   Nothing can equal the artleffnefs of this compofition.   The tone of color and the ftyle of drawing are in admirable harmony, and that harmony could create nothing purer or more divine than the form of the Infant Jefus, and the feeling of adoration of the little Saint John."

The Old Masters.

LANDOR'S IMAGINARY CONVERSATIONS.

[CARDINAL ALBONI.]

TITIAN ennobled men; Correggio raifed children to angels; RAPHAEL performed the more arduous work of reftoring to woman her priftine purity. Perugino was worthy of leading him by the hand. I am not furprifed that Rubens is the prime favorite of tulip-fanciers; but give me the clear warm mornings of Correggio, which his large-eyed angels, juft in puberty, fo enjoy. Give me the glowing afternoons of Titian; his majeftic men, his gorgeous women, and (with a prayer to protect my virtue) his Bacchantes. Yet, Signors! we may defcant on grace and majefty as we will, believe me, there is neither majefty fo calm, concentrated, fublime, and felf-pos-

11

seffed (true attributes of the divine); nor is there grace at one time fo human, at another time fo fuperhuman, as in RAPHAEL.

He leads us into heaven; but neither in fatin robes nor with ruddy faces. He excludes the glare of light from the fanctuary; but there is an ever-burning lamp, an ever-afcending hymn; and the purified eye fees, as diftinctly as is lawful, the divinity of the place.

I delight in Titian; I love Correggio; I wonder at the vaftnefs of Michael-Angelo; I admire, love, wonder, and then fall down before RAPHAEL.

La Vierge a la Redemption.

THIS Madonna, which we were induced to include in the prefent collection, by the twofold claim it poffeffes, of great fweetnefs of expreffion, and a diffimilarity to any other of the number chofen, we are very nearly convinced, from careful inveftigation, was not executed by Raphael, though it paffes generally unqueftioned as fuch. This fact being admitted, critics will find little difficulty in pointing out its defects,—ftiffnefs, want of dignity and character, etc.,—which, however correct the judgment, would hardly have been difcovered, or at leaft, avowed, while the paternity of the original was undoubted. Probably the painting was executed by the immediate fcholars of the great mafter, and may have received fome touches from his own hand.

The original is faid to be in the poffeffion of M. Raphael Tofoni, Profeffor of Chemiftry at Milan.

Letter from Raphael

TO HIS UNCLE.

Written from Rome, July 1st, 1514.

[Probably no apology is required for introducing this letter of Raphael's;—one of the few that have been preserved. If there were, we should urge, first,—in general terms, the vast superiority of original letters over formal Biography, however ingenious and able, in bringing us into actual contact with the subject of our inquiry; and secondly, the unusual interest of the incidents narrated, and the charming simplicity of style, would be sufficient reasons for the republication of the particular letter in question.—Ed.]

DEAR Uncle and Second Father,——I have received a letter from you, to me most gratifying, since I find that you are not angry with me; indeed, you would be wrong to be so, for consider how irksome it is to write when there is nothing important to communicate. But now that there is important matter to talk about, I reply.

In the firft place, with regard to taking a wife, I anfwer that, as to the one you firft intended to give me, I am moft happy, and thank God conftantly that I neither married her nor any other, and in this refpect I have been wifer than you, who wifhed to give her to me.

I am fure you muft now yourfelf be convinced that, had I followed your advice, I fhould not have been in the pofition in which I am. At this moment I find that I have property in Rome to the amount of 3,000 gold ducats, and an income of 50 gold crowns. His Holinefs allows me 300 gold ducats (annually) for fuperintending the building of St. Peter's : this provifion is fecured to me for life.

Other fuch falaries are in profpect, in addition to which I am paid whatever I choofe to afk for my works, and I have begun another room for His Holinefs, which will amount to 1,200 gold ducats ; fo that, dear uncle, I do honor to you and all my relations, and to my native place ; but I ceafe not to hold you in my heart, and when I hear you named, it is as if I heard my father named. Do not, therefore, complain becaufe I do not write ; I might rather complain of you who have always the pen in your hand, and yet fuffer fix months to intervene between one letter and another.

To return to the fubject of the wife, from which I have
digreffed : you are aware that Santa Maria in Portico
(Cardinal Bibiena) wifhes to give me a relation (a grandniece) of his, and, on condition of obtaining your confent, and that of my uncle the prieft, I promifed to do
whatever his Eminence wifhed. I cannot break my
word; we are more than ever ready to conclude the
affair, and I will foon inform you of all. Do not be
offended that this bufinefs thus takes its good courfe;
if it fhould come to nothing, I will then do whatever
you wifh, and know, if Francefco Buffa has good
alliances within his reach, that I can boaft fome too;
for I can find a handfome lafs in Rome, of excellent name, both fhe and hers; her friends, indeed, are
ready to give me a dowry of 3,000 gold crowns with
her. Meanwhile, I live in Rome, where 100 ducats
are more worth having (all things confidered) than 200
in Urbino; of this be fure. With refpect to refiding
in Rome, I can no longer remain elfewhere for any
length of time, on account of the building of St.
Peter's—for I am in Bramante's place : but what place
in the world is more glorious than Rome ? and what
undertaking more honorable than St. Peter's—the firft
temple in the world—the greateft ftructure that has
ever been feen, and which will coft more than a million

of gold? Know that the Pope has determined to spend 60,000 ducats annually for this building; he thinks of nothing else. He has associated with me, in the direction, a very learned friar, more than eighty years old; the Pope sees he cannot live long, and has appointed him as my colleague, as he is a man of great reputation and experience, in order that I may learn from him, if he has any excellent secret in architecture, and that I may become accomplished in this art; he is called Fra Giacondo. Every day the Pope sends for us, and consults with us for a while about this building.

I beg you will go to the Duke and Duchess, and tell them I know they will be pleased to hear that a servant of theirs does himself honor, and commend me to their Highnesses. I commend myself unceasingly to you. Greet all friends, especially Ridolfo, who has so much affection for me.

"EL VOSTRO RAFAEL, Pittore in Roma.

"Alli primo Luglio, 1514."

from "The Palace of Art."

By Tennyson.

*   *   *   *   *   *

OR the maid-mother by a crucifix,
    In tracts of pasture sunny-warm,
Beneath branch-work of costly sardonyx
    Sat smiling, babe in arm.

Or in a clear-walled city on the sea,
    Near gilded organ-pipes, her hair
Wound with white roses, slept St. Cecily;
    An angel looked at her.

Or, thronging all one porch of Paradise,
    A group of Houris bowed to see
The dying Islamite, with hands and eyes
    That said, we wait for thee.

Or mythic Uther's deeply wounded son
  In some fair space of sloping greens
Lay, dozing in the vale of Avalon,
  And watched by weeping queens.

Or hollowing one hand against his ear,
  To list a footfall, ere he saw
The wood-nymph, stayed the Tuscan king to hear
  Of wisdom and of law.

Or over hills with peaky tops engrailed,
  And many a tract of palm and rice,
The throne of Indian Cama slowly failed
  A summer fanned with spice.

Or sweet Europa's mantle blew unclasped
  From off her shoulder backward borne;
From one hand drooped a crocus; one hand grasped
  The mild bull's golden horn.

Or else flushed Ganymede, his rosy thigh
  Half buried in the Eagle's down,
Sole as a flying star shot through the sky
  Above the pillared town.

12

Nor thefe alone : but every legend fair
  Which the fupreme Caucafian mind
Carved out of Nature for itfelf, was there,
  Not lefs than life defigned.

*   *   *   *   *

La Sainte Famille.

THIS fpirited and interefting group was long fuppofed to have been executed exprefly for Francis
I. But fubfequent inveftigation proves it to have
been painted about 1518, to the order of Lorenzo
de Medici, Duke of Urbino, probably for prefentation
to Francis I.

For the fake of diftinction, the picture is known as
the "Benediction,"—from the pofture of one of the
angels,—though it is generally called fimply "The
Holy Family," and, fays M. Guizot, needs no other
defignation. "Perhaps no other of his compofitions,"
he adds, "is fo pure in ftyle, fo lofty and holy in expreffion. All the perfons in the picture are evidently
filled with holy thoughts."

The Infant Jefus is fpringing from his cradle into
the arms of his mother; he is adored by St. John, pre-

fented to him by Saint Elifabeth. An angel is feen
fcattering flowers on the Virgin; another kneels in
homage; and St. Jofeph is abforbed in meditation.

Kugler refers to it as "peculiarly excellent." He
fays, "The whole has a character of cheerfulnefs and
joy; an eafy play of graceful lines, and the nobleft
forms which unite in an intelligible and harmonious
whole."

The original is preferved in the Louvre. Its dimen-
fions are large; it being fix feet eight and a half inches
high, by four feet feven inches wide; the heads are
natural fize.

Painting.

By Prosper M. Wetmore.

Peopling, with art's creative power,
The lonely home, the silent hour.

'TIS to the pencil's magic skill
    Life owes the power, almost divine,
To call back vanished forms at will,
    And bid the grave its prey resign;
Affection's eye again may trace
    The lineaments beloved so well:
'Tis there the childless mother pays
    Her sorrowing soul's idolatry;
There love can find, in after days,
    A talisman to memory.
'Tis thine, o'er History's storied page,
    To shed the halo light of truth;
And bid the scenes of by-gone age
    Still flourish in immortal youth—

The long forgotten battle-field,
    With mailed men to people forth;
In bannered pride, with fpear and fhield,
    To fhow the mighty ones of earth—
To fhadow, from the holy book,
    The images of facred lore;
On Calvary, the dying look
    That told life's agony was o'er—
The joyous hearts, and gliftening eyes,
    When little ones were fuffered near—
The lips that bade the dead arife,
    To dry the widowed mother's tear;
Thefe are the triumphs of the art,
    Conceptions of the mafter-mind;
Time-fhrouded forms to being ftart,
    And wondering rapture fills mankind!

Led by the light of Genius on,
    What vifions open to thy gaze!
'Tis nature all, and art is gone,
    We breathe with them of other days:
Italia's victor leads the war,
    And triumphs o'er the enfanguined plain:
Behold! the Peafant Conqueror
    Piling Marengo with his flain:

That fun of glory beams once more,
 But clouds have dimmed its radiant hue,
The fplendor of its race is o'er,
 It fets in blood on Waterloo!

What fcene of thrilling awe is here!
 No look of joy, no eye for mirth;
With fteeled hearts and brows auftere,
 Their deeds proclaim a nation's birth.
Fame here infcribes for future age,
 A proud memorial of the free;
And ftamps upon her deathlefs page,
 The nobleft theme of hiftory.

To the Virgin.

FROM THE GERMAN OF NOVALIS.

BY RICHARD MONCKTON MILNES.

IN thousand forms, Eternal Maid,
   Has pious Art imagined Thee,
But never wert thou so portrayed,
   As once, that once, Thou cam'st to me.
I only know that since that sight
   I take no thought of night or day,
And all the world's material might
   Flees like a shamed child away.
Thou bad'st me drink, and since full deep
   I drained the cup thy hand had given,
A perfect rest, that was not sleep,
   Passed to my soul, and made it Heaven.

Madonna di San Sisto.

HIS favorite Madonna was painted, according to Vafari, as an altar-piece for the high altar of the church of the Black Friars of San Sifto in Piacenza. It has, however, been fuppofed that it was defigned for a proceffion picture, to which opinion feveral writers of good authority incline.

Above, are the Virgin and the Infant Jefus upon clouds, in a brilliant glory of countlefs angel heads, and below, St. Sixtus, on one fide, and St. Barbara on the other.

"Of all the figures of the Virgin," fays De Quincey, "his genius created, none was conceived in a fuller, and, if we may ufe the term, a more picturefque ftyle." "We muft further," he adds, "point out to admiration the two cherubim at the foot of the compofition—marvels of color, beauty, expreffion, and life, which

13

abfolutely feem coming out of the canvas, fuch falient relief has the painter given them."

Kugler declares this Madonna to be "one of the moft wonderful creations of Raphael's pencil; fhe is at once the exalted and bleffed woman of whom the Saviour was born, and the tender earthly Virgin whofe pure and humble nature was efteemed worthy of fo great a deftiny." * * * "This is a rare example of a picture of Raphael's later time, executed entirely by his own hand. No defign, no ftudy of the fubject for the guidance of a fcholar, no old engraving after fuch a ftudy, has ever come to light. The execution itfelf evidently fhows that the picture was painted without any fuch preparation."

This marvellous picture now forms the gem of the Royal Gallery at Drefden, which holds the firft rank among all the collections in Germany. It was commenced by Auguftus II., King of Poland; and this painting was purchafed for the Gallery by Auguftus III., for 22,000 crowns.

Mrs. Jamefon writes, "Six times have I vifited the city made glorious by the poffeffion of this treafure, and as often, when again at a diftance, with recollection difturbed by feeble copies and prints, I have begun to think, 'Is it fo indeed? is fhe indeed fo divine? or

does the imagination encircle her with a halo of religion and poetry, and lend a grace which is not really
there?' and as often, when returned, I have ftood
before it and confeffed that there is more in that form
and face than I had ever yet conceived.

"In the fame Gallery is the lovely Madonna of the
Meyer family; inexpreffibly touching and perfeét in its
way, but conveying only one of the attributes of Mary,
her benign pity, while the Madonna di San Sifto is an
abftraét of all."

A modern traveller in Europe, a fcholar and a man
of cultivated tafte and refined fenfibilities, thus records
the impreffion made upon him by this fublime compofition:—"The fpeétator feels, at firft, a little curious
and puzzled to account for its effeéts; for this aftonifh-
ing piéture does not feem to have been elaborated with
the patient pencil that has wrought fo unweariedly
upon many other famous fubjeéts, but rather to have
been thrown off, almoft as though it had been in
water-colors, by an infpiration of divine genius, in a
fudden jubilee of its folemn exercife, with a motion
of the hand, at the laft height and acme of its attain-
ment. * * * Never before by any like produétion
had I been quite abafhed or overcome. I could except
to, and ftudy and compare, other piétures: this paffed

my underftanding. Long did I infpect, and often did
I go back to re-examine this myftery, which fo foiled
my criticifm, and conftrained my wonder, and convinced me, as nothing vifible befides had ever done,
that if no picture is to be worfhipped, fomething is to
be worfhipped; that is to be worfhipped which fuch a
picture indicates or portrays. But the problem was too
much for my folving. I can only fay, it mixed for me
the tranfport of wonder, with the ecftafy of delight;
it affected me like the fign of a miracle; it was the
fupernatural put into color and form; for certainly no
one, who received the fuggeftion of thofe features, the
fenfe of thofe meek, fubduing eyes, could doubt any
longer, if he had ever once doubted, of there being a
God, a heaven, and, both before and beyond the fepulchre, an immortal life. No one who caught the
fupernal expreffion of the whole countenance, could
believe it was made of matter, born of mortality, had
its firft beginning in the cradle, or could be laid away
in the grave, but rather was of a quite datelefs and
everlafting tenure. I would be free even to declare,
that, in the light which played between thofe lips and
lids, was Chriftianity itfelf,—Chriftianity in miniature,
for the fmallnefs of the fpace, I might incline to ex-
prefs it, but that I fhould query in what larger prefent-

ment I had ever beheld Chriftianity fo great. Mont Blanc may fall out of the memory, and the Pafs of the Stelvis fade away; but the argument for religion,— argument I call it,—which was offered to my mind in the great Madonna of Raphael, cannot fade."*

We cannot more fitly clofe this fketch, than by the following invocation by Wordfworth :—

> " Mother! whofe virgin bofom was uncroft
> With the leaft fhade of thought to fin allied!
> Woman! above all women glorified;
> Our tainted nature's folitary boaft;
> Purer than foam on central ocean toft;
> Brighter than eaftern fkies at day-break ftrewn
> With fancied rofes, than the unblemifhed moon
> Before her wane begins on heaven's blue coaft,
> Thy Image falls to earth. Yet fome, I ween,
> Not unforgiven, the fuppliant knee might bend,
> As to a vifible Power, in which did blend
> All that was mixed and reconciled in thee,
> Of mother's love with maiden purity,
> Of high with low, celeftial with terrene."

----

* Pictures of Europe.

To the Genius of Art.

BY ESTELLE ANNA LEWIS.

THOU art a beam from God——the brighteſt ray
  That heaven hath earthward ſent to cheer the
    ſoul
And animate it in its houſe of clay,
  With dreams of light, and life, and glory's goal.
Here, mutely worſhipping, I gaze on thee,
  Till naſcent haloes dawn around thy brow,
And from the portals of eternity,
  The laurelled dead, returning, round thee bow.
There, bent o'er Fornarina's ſainted face,
  Feeding his ſoul, eternal RAPHAEL kneels,
As if in its pale hues he ſtill can trace
  Beauty, ſurpaſſing all that Heaven reveals;
Angelo——Titian——all the immortal great,
·Glide in, and at thy feet for inſpiration wait.

The Marriage of Joseph and Mary.

"WHEN Mary was fourteen years old, the prieſt Zacharius inquired of the Lord concerning her, what was right to be done; and an angel came to him and ſaid, 'Go forth and call together all the widowers among the people, and let each bring his rod (or wand) in his hand, and he to whom the Lord ſhall ſhow a ſign, let him be the husband of Mary.' And Zacharias did as the angel commanded, and made proclamation accordingly.

"And Joſeph the carpenter, a righteous man, throwing down his axe, and taking his ſtaff in his hand, ran out with the reſt. When he appeared before the prieſt, and preſented his rod, lo! a dove iſſued out of it—a dove dazzling white as the ſnow,—and after

fettling on his head, flew towards heaven.   Then the
high prieft faid to him, 'Thou art the perfon chofen
to take the Virgin of the Lord, and to keep her for
him.'   And Jofeph was at firft afraid, and drew back,
but afterwards he took her home to his houfe, and faid
to her, 'Behold, I have taken thee from the temple
of the Lord, and now I leave thee in my houfe, for I
muft go and follow my trade of building.   I will return to thee, and meanwhile the Lord be with thee
and watch over thee.'

    "So Jofeph left her, and Mary remained in her
houfe."

www.ingramcontent.com/pod-product-compliance
Lightning Source LLC
Chambersburg PA
CBHW021813190326
41518CB00007B/582